WHAT DO YOU BELIEVE?

(Numbers Don't Lie or Do They?)

J. STUART FLEMING

WHAT DO YOU BELIEVE © 2013 was published
in cooperation with CreateSpace (createspace.com)
ISBN-13 978-1491076200

Table of Contents

What Do You Believe?
(Numbers don't lie or do they?)

Introduction

Caffeine is bad for you. "Caffeine can cause a short, but dramatic increase in your blood pressure", (Mayo Clinic, 10/21/2011). "Caffeine in reasonable amounts is beneficial." A new study has shown that "caffeine boosts power in older muscles, suggesting the stimulant could aid elderly people to maintain their strength, reducing the incidence of falls and injuries" (Society for Experimental Biology 6/30/2012).

Alcohol is bad for you. "Alcohol can produce detectable impairments in memory after only a few drinks", (NIAAA, 10/2004). One or two alcoholic drinks a day may be beneficial. Two new studies show "a bit of alcohol each week might just stave off two painful conditions some women face: rheumatoid arthritis and osteoporosis" (CBS News, 7/11/2012).

Human activity is causing global warming. "An increasing body of science points to rising dangers from the ongoing buildup of human-related greenhouse gases" (NY Times, 2/22/2013). The earth is actually entering a period of climatic cooling. "Natural climate cycles have already turned from warming to cooling, global temperatures have already been declining for more than 10 years, and global temperatures will continue to decline for another two decades or more" (Forbes, 5/31/2012).

Dietary cholesterol is a primary cause of heart disease. "People who have high blood cholesterol have a greater chance of getting coronary heart disease, also called coronary artery disease" (National Institutes of Health, 9/19/2012). Dietary cholesterol is not a primary cause of heart disease. "There is no conclusive evidence that either cholesterol in the diet or plasma cholesterol causes heart disease." "It's not even really a good predictor of heart disease. Half the people who are admitted to hospitals in America with cardiovascular disease have normal cholesterol – and half the people with elevated cholesterol have normal hearts" (Fox News, 2/15/2013).

What do you believe? These are but a few examples of the many inconsistencies that we might encounter as we attempt to understand a multitude of issues. Often these issues originate in the scientific literature and are then brought to public attention by the news media or other popular outlets. In each of the above cases, the initial statement is the "conventional wisdom", i.e., most likely believed by the majority of people including a number of highly qualified scientists. However, the conflicting opinions likewise have substantial support, also including highly qualified scientists. How can there be two such contrary opinions? In particular, how can we navigate through these apparently incongruent findings and opinions to make rational choices about issues of health, finance and other important aspects of life?

You may note that while there appear to be inconsistencies in some of the above examples, these may not necessarily all be inconsistencies but rather reflections of differences in individual responses and or perceptions. In the case of examples such as caffeine and alcohol, it is perfectly possible for both positive and negative consequences of their use to occur. Which outcome dominates in any particular individual likely depends on genetics and variations that can occur in one's particular biochemical and physiological makeup. For example, caffeine or alcohol in modest amounts might be beneficial to you but harmful to me. The recent fish oil controversy described below further dramatizes this point.

Howard LeWine, M.D., Chief Medical Editor Internet Publishing, Harvard Health Publications (7/12/2013) has nicely summarized the fish oil question. Because fish oil contains a high concentration of omega-3 fatty acids, fish oil supplements have been promoted in recent years as an easy way to protect against deficiencies in these substances. Dr. LeWine explains that omega-3 fatty acids play important roles in a variety of normal development and functional areas. Deficiencies have been linked to a variety of health issues including cardiovascular disease and arthritis among others. As a result, fish oil supplements have often been included in daily vitamin and supplement regimens. However, a recent study reported by scientists at the Fred Hutchinson Cancer Research Center in Seattle (Journal of The National Cancer Institute, 7/13/2013) found a substantial increased risk of prostate cancer in men eating a lot of fish or taking potent fish oil supplements.

If you happen to be a middle age male with a family history of cardiovascular disease and you have been taking a daily fish oil supplement thinking that it may be beneficial in preventing cardiovascular disease, what do you do now - stop taking the fish oil for fear of prostate cancer or continue it considering your family history? With only the above information, which is what most people would have, either choice seems to be a gamble. In order to arrive at a more informed conclusion, one would have to delve into the specifics of the scientific studies that support each opinion. Armed with a capability to assess the studies and reanalyze the data, one may be able to determine which opinion has the strongest base of supporting data. Unfortunately, most of us don't have the background or opportunity to perform such an analysis and we defer to our physician's opinion, which may or may not be well informed. In any case we need to form an opinion and plan of action that is based on a rational judgment of our own particular situation and all the related information available to us.

The fish oil example is representative of many similar health-related questions that we encounter on an almost daily basis. While there are usually no easy answers short of the detailed analysis referred above, armed with insight into some of the common pitfalls encountered in scientific studies and data analysis, the average person can often achieve a more well informed perspective on these issues. A number of the most common pitfalls will be explained and explored in the chapters to follow.

Humans have reached the point in evolution where we possess the mental capacity to accumulate information and, over time, build an ever-expanding base of knowledge. However, in order to build a reliable knowledge base, we need to successfully harvest those kernels of valid information from the vast sea of information we are immersed in, much of which is misinformation. This can be far more difficult than it might appear on the surface, even for highly trained scientists or experts in other fields. Hopefully, concepts to be discussed in the following pages will be beneficial in addressing this challenging task.

Ideally, one could obtain a laymen's guide for critically evaluating scientific studies. Such a guide might allow the average, non-scientist, to ask the critical questions and to seek out key pieces of information. This may then lead to an informed opinion regarding the appropriateness of the conclusions reached and presented to the public. It's surprising how many supposedly well-run scientific studies reach poorly supported conclusions. The reasons for this are many and varied and some will be discussed in the pages to follow.

In addition to considering common pitfalls of scientific studies and poorly supported conclusions, we need to consider the broader topic of critically evaluating all information we are presented with. We need to hone our skills in distinguishing between valid information and misinformation. We need to be aware of the essential mental steps that we must go through when presented with new information in order to best determine which category, valid information or misinformation, it belongs to. Unfortunately, much information presented to us is more often than not misinformation, regardless of its source. That may sound overly skeptical but that old saying, "don't believe anything you hear and only half of what you see", may be good advice at the outset. Putting it another way, how do we know when information presented to us reflects reality?

Now that introduces another blockbuster and key piece of the puzzle. What do we mean when we refer to reality? How do we know when we are dealing with reality? In fact, what is reality and is there any such thing as absolute reality?

Chapter 1
Reality

In attempting to define reality, one is led to a laundry list of terms or concepts that actually get no closer to an understanding of the nature of reality than the word we began with, reality itself. We might be told that reality is simply that which actually exists, a concept that isn't much better than the simple word "reality". How can we ever be sure of what is real or what actually exists? On the surface one may think that this question is easily answered but, as discussed in **"A View From The Rabbit Hole"**, current theories of relativity and quantum physics introduce complexities that are very difficult to reconcile.

One possible construct of reality, and the one chosen to be adopted here, is that there are two spheres of reality, subjective reality that originates in the human mind and absolute reality that exists in nature and is independent of human presence. Thus, absolute reality is universal while subjective reality may be valid for only a given individual or a segment of the population but is probably not universally accepted by everyone.

Some may question whether we can ever be sure that absolute reality actually exists. In fact, one school of thought is that since all concepts of reality are formulated in the human brain, all reality is simply a composite of subjective realities.

This idea is extensively explored in a book by Robert Lanza and Bob Berman, "Biocentrism" (BenBella Books, Inc., Dallas, TX, 2009). Lanza is a highly respected scientist who some have compared to the genius of Albert Einstein. Biocentrism, as detailed in Lanza's book, takes the concept of there being no absolute reality external to human consciousness to the ultimate conclusion, i.e., that everything in the universe is structured and brought to a state of reality in the human mind. While this, on the surface, sounds like an extreme, maybe even crazy, view of reality, his ideas are formulated from an impressive background in both the physical and biological sciences.

While Lanza's rationale is compelling, it would seem difficult to reconcile it with the fact that multiple individuals often make consistent observations of the natural world. The five human senses lead to a common impression of the environment that is typically shared by many. For example, ten people individually viewing a scenic wonder such as The Grand Canyon or Niagara Falls will most likely provide nearly identical descriptions of what they saw. In order for the image of Niagara Falls to be totally formulated within the human brain without Niagara Falls existing in the realm of absolute reality, some strange theory of mental entanglement would have to be invoked where once a first image is formulated by one human brain, all successive images would simply be dependent exact copies. It seems that such a totally human-dependent concept of reality is counterintuitive and highly unlikely. Thus, for the purposes of our present discussion, the prerogative will be taken to adopt the view of their being two compartments of reality, absolute or universal reality and subjective or individual reality.

Absolute reality is revealed via a systematic rational process for observing and documenting natural phenomena, i.e., a scientific method. This is obvious for many, readily observed, macroscopic features of the universe, for example the existence of the oceans, the moon and the planets of our solar system. However, it may also be true of intellectually derived features that have not been directly perceived via the five senses. For example, no one has ever seen the earth's core. However, a scientifically based concept suggests that it is a sphere of molten iron-nickel alloy. This widely accepted notion is, in effect, a part of our collective absolute reality.

Many would also consider certain philosophical matters, e.g., concepts of heaven and God to be components of an absolute reality. The appropriateness of this conclusion can be argued long and hard by individuals on both sides of the issue. On the one hand, most cultures from the time of early man to this day have believed in some form of spiritual presence or deity responsible for creating the universe and everything in it, including all of us. For the vast majority of humans who have ever walked on earth, the existence of a God or Gods has been a central component of absolute reality. On the other hand, there is not a single shred of hard scientific evidence to support such a belief.

While many pages could be devoted to pursing the appropriate categorization of God and Religion, an arbitrary convention will be invoked at this juncture in order to allow us to move on to the more mundane topics we wish to address here. The convention is based on acceptance of a practical point of view. For present purposes we will assume that all ideas, knowledge and beliefs of a philosophical nature are best accommodated in the sphere of subjective reality. While they may also be components of the broader sphere of absolute reality, we can never be sure of that. Regardless of how we choose to categorize our philosophical thoughts and beliefs, they are reality to us. Something cannot be semi-real. **If we believe something, we believe it. Furthermore, it should not be critically important to us whether or not others believe it. In fact, the world would be a much better place, if we didn't insist on others accepting our**

particular personal beliefs, i.e. our own particular subjective realities. To state it directly, it is being proposed that all matters of philosophy, including religious and political philosophy, should not be regarded as matters of absolute reality but matters of individual subjective realities. The religious conflict we see in the world today that is causing so much suffering and hatred would probably not exist if everyone accepted the idea that one's own religious beliefs were part of their own subjective reality rather than insisting that they are part of an absolute reality that needs to be embraced by all.

For the most part, we can never be sure that a philosophical belief belongs in the category of absolute reality. Some, such as Dr. Lanza, believe that no absolute reality at all actually exists. However, the concept of there being two compartments of reality, an absolute reality concerning the physical universe as revealed through scientific study and a subjective reality created by the human mind, would seem to be more in tune with conventional thought.

Regardless of how we choose to categorize reality, it's vitally important that we make full use of our intellect in formulating our beliefs. We need to be diligent in verifying our beliefs and constantly checking them to be sure they are congruent with the widely accepted base of prior knowledge.

The scientific study of nature may be the most productive route for gaining a glimpse of absolute reality. Ideas and beliefs formulated through a process of critical analysis of study observations and data combined with intellectual honesty are key components of such an approach. One could imagine that there is a broad range of human insight with respect to absolute reality. It may be that only a small percentage of the average human portfolio of ideas and beliefs can be regarded as part of the spectrum of absolute reality while the larger portion, e.g., 90% or more of that portfolio must be considered subjective reality. In the case of the most educated and critically analytical of the human species, achieving a portfolio of ideas and beliefs that approaches a significantly greater absolute reality content, e.g., 50% or more might either be common, very rare or even essentially impossible. We may never know. The above percentage examples, while arbitrarily selected, simply make the point that many people probably function primarily in the realm of subjective reality. For any who truly wish to gain a greater glimpse of absolute reality, continually seeking education accompanied by critical analysis and intellectual honesty is essential. It is from this point of view that we undertake an effort to differentiate between opposing concepts, beliefs and points of view in order to incorporate the most intellectually appropriate into our own portfolio of ideas and beliefs.

As pointed out above, most of us are possibly in touch with only a small bit of the absolute component in our total realm of reality, most of it being composed of ideas and opinions of a subjective nature. To really know or understand something in the scientific sense allows us to gain a glimpse of absolute reality while to think we know or understand something usually resides in the realm of subjective reality.

A wise person once noted that one does not begin to become educated until he realizes how little he knows. Considering the entire world population of approximately 7 billion people, it's been estimated that only a small fraction, 10% or less, are scientifically literate. To be considered scientifically literate, an individual should be able to satisfactorily answer a standard set of questions indicating a basal level of understanding of key scientific areas such as chemistry, physics and biology. In fact it has been estimated that only about 28% of the U.S. population is scientifically literate. This is not to imply that those who are not scientifically literate are stupid. To the contrary, there are many who, although being scientifically illiterate, may display a genius associated with other intellectual or artistic aspects of life. Likewise, some who are scientifically literate may be illiterate in other non-scientific areas.

Thus, it is not intended to demean those of limited scientific literacy and who, presumably primarily live in the realm of subjective reality. To the contrary, the universe of subjective realities is where we spend all or most of our lives. It contains most of what matters to us in everyday life and our skills in navigating the multitude of subjective realities usually determines how successful we will be in conducting our own lives.

We may never understand the fundamental nature of matter or the significance of the warping of space-time but does that really matter to us in our day-to-day lives? Is a firm grasp of particle physics or cosmology essential and relevant to everyday life? I think most of us are comfortable leaving true knowledge of most aspects of absolute reality to that small group of highly educated geniuses who communicate via complex mathematical equations. Hopefully they will eventually translate their revelations into a language the average person can comprehend, thus, providing us with a small glimpse of the elusive realm of absolute reality.

The important thing for most of us to master is an ability to sort through the vast amount of information we are presented with and determine the probability that any particular piece of information relates to absolute reality.

We are presented with a deluge of information on a daily basis, much of it being represented as scientifically determined and presumably a component of absolute reality. Many times those presenting such information know full well that it is likely not accurate or that it is out right fraudulent. This is often done to sell us something, e.g., a magnetic bracelet that has none of the special powers represented, a diet that is sure to result in significant weight loss but, more often than not, ends up on the trash heap of previously failed diets or a political position that is sold to us as a benefit but ends up only benefiting the politician doing the selling. We need to be able to evaluate this deluge of information and weed out the vast majority of it that is actually misinformation.

The first, and perhaps most important, bit of advice is to question everything we hear and read. Accept nothing as valid information until we are convinced of its unassailable truth. Torture every cause and effect we are presented with by asking, what else could explain the result. While we may be tempted to readily accept findings that agree with our preconceived ideas or wishes, we should be particularly cautious. We should question everything and be convinced of its truth before accepting such information as valid.

Unfortunately, there is no magic formula for determining which of two opposing viewpoints is most accurate and important to us or if, in fact, either is relevant. Even two highly trained scientists might disagree as to their accuracy and relevance. This being the case, what chance does the average person have in forming a valid opinion? While we may not be able to arrive at a conclusion that is correct in the purest sense, i.e., correct beyond all reasonable doubt, it is possible to become familiar with several of the most common analytical traps and biases that can lead to unjustified conclusions. Unfortunately, it is such unjustified conclusions that often find their way into the public domain and become components of the "conventional wisdom". By recognizing the possible presence of analytical traps and biases, we can question conclusions as presented and be guarded in our acceptance of them.

While most investigators are forthright and honest, some are frankly dishonest and willing to cheat to advance personal interests (John Timmer; Ars Technica, 10/1/2012 & Geoff Maslen; University World News, 4/25/2013). However, in most cases where unjustified conclusions are reached, it's more a case of investigators falling into unrecognized analytical traps or simply succumbing to a bit of over exuberance rather than frank dishonesty

In the chapters to follow, we will consider some of the primary factors that can lead to inaccurate conclusions or opposing conclusions by two different people looking at the same set or similar sets of study results.

The goal will be to equip ourselves with, not so much the ability to analyze scientific data, but to understand and recognize potential unintended traps and other causes of misinterpretation of study results.

Note! Perhaps the most obvious category of opposing opinions occurs in the realm of political subjective reality.

This "non-scientific" category will be commented on later but for now, we will only focus on results derived from scientific studies.

As we go through life, we each build an ever-growing complement of ideas and beliefs as a result of our particular experience and education. It is this complement of ideas and beliefs that comprise our own version of reality. This subjective reality or individual reality may or may not be variably congruent with the existing absolute reality.

While subjective reality is dependent on the human species and human mental activity, absolute reality would have no such dependence and presumably would exist in the total absence of all human life. Certain individual subjective realities might be common among particular groups of humans such as families, tribes, certain cultures or perhaps, among nearly all humans. However, the widespread extent of collective beliefs alone does not necessarily mean that such beliefs are components of absolute reality. For example, until quite recently atoms were considered to be the most elemental particles of matter and were the basic building blocks of everything around us from soup to nuts, so to speak. At the time, this was a component of the subjective realities of nearly every educated person and, furthermore, firmly believed to be a component of absolute reality. Not so!

The physicists now tell us that atoms themselves are composed of even smaller units, which in turn are composed of still smaller units, and that if the sequence is followed far enough, we are told that the particle nature of the units disappears leaving us with wave-like ripples in space. Our reference and our orientation that we have depended on to understand what the universe is and what we are has evaporated in the wake of modern quantum theory and we are left with the statement that everything we see and touch, including each of us, is basically empty space occupied primarily by the four fundamental forces. It's no wonder that such revelations can shake to the core our confidence in the existence of any form of absolute reality.

The traditional religious among us might claim that the only element of absolute reality that we really need is the existence of God. While that may well be true, the absolute existence of God must be taken as a matter of faith. Our model suggests that the concept of God is more rationally placed in the realm of subjective reality. One problem with considering God to be a component of absolute reality is that there are many variations in religions and the nature of God. The modern monotheistic concept of God is inconsistent with there being many variations in both the God concept and God's wishes of how we humans are to lead our lives. For example, there are clearly huge discrepancies in the teachings of Islamic fundamentalism, a very casual form of Christianity such as Unitarianism, and Eastern religions such as Buddhism and Hinduism. It would seem to make much better sense to regard both religion and the God concept to be products of the human mind and, as such, components of various subjective realities.

Regardless of whether a person's beliefs are components of subjective, absolute or both categories of reality, they are his reality and deserve to be respected. In the final analysis, reality is reality, not semi-reality or quasi-reality. A person's reality is critically important to him and provides the foundation for his journey through life.

However, having said that, we need to also acknowledge that our personal structure of reality will likely change as we mature, gain in experience and become educated. We should be open to constantly examining our beliefs and if rational thought suggests that certain beliefs should be modified, we should recognize that, although possibly painful, modification will usually lead us to a more satisfying outlook on our lives.

It is with this perspective on reality and our own complement of beliefs, that the chapters to follow will hopefully aid in bringing our individual beliefs and, thus, our own subjective realities closer to the realm of absolute reality.

Chapter 2

What is Science?

While most of us may recognize that science is the systematic study of nature we may not be fully aware of the essential characteristics of an investigation that qualify it as a scientific study? There are many concepts, ideas, opinions and postulates that pose as having a scientific basis but, in fact, a number of these have no such basis. We are bombarded by pseudo-science on a daily basis in all forms of the media from TV ads to Internet pitches and solicitations in the mail. "Quasi-scientific" descriptions and claims are commonplace in advertising as well as in a variety of efforts aimed at influencing everything from our individual behaviors to overall public opinion.

The definition of science that we will use here is the following. Science is the body of knowledge of the physical and natural world revealed by means of a systematic study process, i.e., a "scientific method". While there is no one strictly specified method, for any process to be considered a scientific method, it should contain certain essential steps. The essential steps include observation, measurement and experimentation. In the process, scientific knowledge is gradually revealed through formulation, testing and modification of hypotheses. In simplest terms, observing and measuring natural phenomena may often suggest a hypothesis or rudimentary theory. The hypothesis is then tested through experimentation. The most robust experimental testing occurs by means of well-controlled experiments where all variables other than the one being tested are held constant. Once formulated, every effort should be made to disprove the new hypothesis through the experimental sequence. Failure to do so after exhaustive testing establishes the hypothesis as a scientific theory. If the theory survives for an extended period of time, it may achieve the exalted level of a "law of nature".

Turning now to the human element in the process, the scientist. Most importantly, it needs to be pointed out that intellectual honesty is essential if the primary goal of the scientist is to reveal true knowledge of the natural world. Having this in mind, the intellectually honest scientist will torture the hypothesis in every way imaginable in attempt to expose potential weaknesses. If neither the primary scientist nor any colleagues in the scientific community are able to detect weakness in the hypothesis it will likely be nominated as a new scientific fact. Unfortunately, much of what is advanced in the public domain as true science has not met these critical standards and needs to be regarded as "quasi-science" and, in fact, as previously pointed out, some may be outright fraud.

Scientific Data

There are a variety of ways to gather data and the study design determines the manner in which data are collected and analyzed. Skill and care in designing a study can play a major role in determining the accuracy and reliability of conclusions reached. One needs to consider the type of study used to generate the data. Have the data come from surveys or data mining of past events or have the data been generated de novo as a product of a newly designed scientific study? Have sufficient data been collected to warrant or support a conclusion? These and similar questions concerning study design comprise the initial level of evaluation necessary in addressing the probable validity of scientific conclusions presented to us.

With a basic understanding of the study design and objective, one can then consider details of how the investigator conducted the study. It is at this stage where the informed reader may recognize potential "traps" or other analytical anomalies that could possibly distort study conclusions.

Our discussion here is organized from the standpoint of some of the more common design, methodological, and analytical "traps" that can lead to distorted conclusions. Several real life examples will be used whenever possible to illustrate the manner in which the "trap" can distort conclusions.

Before delving into the mechanics of science, we also need to once again remind ourselves of the entire premise of scientific investigation on a more basic level. The essential goal of every scientific study is to expand the boundaries of our knowledge of nature. That implies that there is a vast compendium of absolute reality and that our present study is aimed at shedding light on an, as yet, unrevealed tiny bit of that reality. It is from the standpoint of this lofty objective that all scientific investigation should precede by following the highest standards of ethics and honesty.

Chapter 3
Science, Conventional Wisdom and Reality

We'll begin by considering an example of a "conventional wisdom" that has become a component of absolute reality for many but that may not warrant inclusion in that exalted category. Everyone with children knows that sugary treats consumed at birthday parties leaves children "hyper" i.e., over stimulated with difficulty in calming down. We know that from personal experience and, besides, its general knowledge, i.e., conventional wisdom. This quasi-scientific fact has probably become a part of our society's absolute reality. The only problem is that this example has no basis in scientific fact. No well-conducted scientific studies have been able to demonstrate a cause and effect relationship between sugar consumption and stimulation or hyper activity in children or anyone else for that matter. Mina Dulcan, M.D., head of child and adolescent psychiatry at Children's Memorial Hospital in Chicago states that "there is elegant research demonstrating that sugar is not at all related to inattention or hyperactivity" and Steven Pliszka, MD, professor of psychiatry,

University of Texas Health Science Center at San Antonio, suggests that the "sugar myth" could probably be expanded to include the idea that food in general has no direct effect on behavior (MedicineNet.com 6/24/2013).

While many may disagree with Dr. Dulcan's point about sugar and hyper activity in children, we, at least, have to acknowledge that such a connection may not be a part of an absolute reality. However, personal experience might be sufficient for us to include it as a part of our own personal, subjective, reality. With respect to Dr. Pliszka's point, we might want to exercise a bit of caution. Sweeping generalized statements should be a warning to us to be on guard for the presence of an analytical trap or bias that could lead to an unjustified conclusion. In assessing these kinds of statements, we should recognize that there could be exceptions to just about everything. While it may be true that most foods in the majority of cases don't affect behavior, certain individuals may be unusually sensitive to particular foods. Without even considering food allergies, some specific foods can have an exaggerated effect on certain people. One that comes to mind is the belief that some people may be particularly sensitive to the L-tryptophan found in turkey and after a Thanksgiving feast tend to become sleepy and lethargic. While this may also be mostly myth, there is some element of truth to the idea that certain meals can leave you drowsy. Thus, we should, perhaps, always be suspicious of sweeping generalities in assessing scientific opinions or, for that matter, any other stated opinions.

At this point, it might be useful to propose a model of reality that may be useful in thinking about the interplay between absolute reality and subjective reality. Some such as Dr. Lanza might argue that there is only subjective reality because all concepts of what there is arise from the human brain and is based on human sensory and intellectual input. However, it might be suggested that there could be, although we will likely never be able to prove it, features of existence that would persist and continue to exist in the total absence of all forms of intelligence. While human perception and intelligence may only be able to sample a small portion of the base of absolute reality, it does provide the foundation upon which we build our subjective realities. While the subject of reality has been previously discussed and could be expanded greatly including a vast reference literature on the subject, the above is hopefully sufficient for our present task of examining the validity of scientific opinion.

Chapter 4
Study Design - Retrospective Studies

While various study design types have been described, we will speak of them as either retrospective studies or prospective studies. Retrospective studies deal with unplanned current events or events in the past while prospective studies are essentially newly designed and conducted experiments. Most human studies are of the former type. In the retrospective study, investigators collect data primarily by observing nature in action. They attempt to discover correlations either via direct observation or through the use of surveys. It's particularly important to realize that in such studies, investigators seldom if ever have the opportunity to establish cause and affect relationships. For example, the objective of such a study might be to see if heart attacks occur more frequently in populations with high levels of plasma cholesterol. If a positive correlation is found, it could suggest, but not prove, that high cholesterol levels lead to heart attack.

Retrospective studies seldom if ever establish a cause and effect relationship. In order to establish a cause and effect relationship, a series of prospective studies, i.e., well-controlled scientific experiments would usually need to be conducted. In such studies, the investigators would develop a tightly controlled protocol so that all variables other than plasma cholesterol, the independent variable, are held constant. Plasma cholesterol would then be intentionally varied in a prescribed way to determine what, if any, affect it has on incidence of heart attack, the dependent variable. Such a prospective study design would allow the investigator to isolate and study the possible cause and effect relationship between plasma cholesterol and heart attack.

Prospective studies are usually blinded i.e., neither the subject nor investigator is aware of the subject's treatment which is usually administered by a third party. In fact In a blinded study, none of those involved, i.e., investigators, subjects or evaluators are aware of study details or data, other than what the experimental protocol allows, usually the minimum amount of information needed to properly conduct and monitor the study. This condition is necessary to avoid human bias in producing the data. Additionally, it is important to be sure that the study is properly powered. That means that the study groups should be large enough so that statistically meaningful conclusions can be drawn considering the likely variability and magnitude of the anticipated outcome. This simply means that the smaller the difference the investigator anticipates seeing between a test group and a control group, the larger these groups will have to be in order for that small difference to achieve the level of statistically significance.

Unfortunately, for both ethical and practical reasons, conducting such well-controlled, properly powered, prospective studies in humans is often not possible. Thus, animal models of the human situation are frequently used as surrogate experiments. Herein lies one of the major "traps" in attempting to reach a useful conclusion relevant to humans. Laboratory animals are not humans and results obtained from animal studies can only suggest but, again, never prove what may occur in man. We'll return to this topic a bit later when we examine pharmaceutical research and the role of animal testing in new drug discovery and development.

Retrospective studies can often be surveys where subjects are questioned about past behavior, e.g., diet, smoking, exercise, etc. The conclusions of such studies are often attempts to correlate health status with various aspects of life style. Observational studies are also often conducted where the investigator simply observes natural phenomena or human activity without intervening other than to take notes. After observing enough such activity, the investigator will review his notes in an attempt to uncover recurring trends that may suggest significant relationships. Jane Goodall is famous for her many observational studies of chimpanzee behavior in the wild. Her work forms the basis of what we currently know about social interactions in chimpanzee families and culture. Many times observational studies are more descriptive of nature as opposed to the goal of firmly establishing cause and effect relationships. However, there have been a number of observational studies of human behavior, which frequently combine questionnaire surveys with observation in attempt to suggest cause and affect relationships. Observations and surveys of life style are often aimed at revealing possible causes of certain important outcomes, e.g., causes of disease and human longevity, subjects of interest to all of us. This type of study will be a major focus for us since the vast majority of human science reported in the press results from observational studies rather than prospective experiments.

Correlations:

The first "quicksand" that we may encounter in evaluating retrospective data has to do with correlations, the most likely analytical result coming from such studies. Our diet, i.e., what we eat, has long been a prime target of survey and observational studies and an area of countless conflicting results, some of which were mentioned at the outset in the Introduction. Often, correlations are sought between a particular human health issue and dietary factors. These correlations, when found, reportedly predict what foods are "good for you" and what foods are "bad for you".

Other examples of observational studies might include studies to find correlations between weather patterns and crop yields, studies looking for correlations between river water temperature and fish breeding habits, as well as countless other combinations. There have been literally thousands of observational studies conducted that provide insight into many aspects of human life as well as many other aspects of the natural world. Survey studies are just another type of observational study except that subjects provide information in response to questions asked by the investigator rather than information being collected directly by the investigator through personal observation. For example, subjects may be asked questions such as, on average, how many alcoholic drinks they consume per week, how many cigarettes they smoke per day or per week or how frequently they consume particular foods. Survey studies can be somewhat less reliable than direct observation studies simply because another person, the interviewee, participates in the information gathering sequence and the interviewee may not be totally honest or accurate in his or her responses

The greatest portion of the information available to us has come from observational studies. In fact, all scientific knowledge begins its journey by progressing from a curiosity to a scientific fact through observations. The correlations observed in these studies provide the foundation for hypotheses and the subsequent design of scientific experiments that will hopefully give rise to higher level cause and effect facts and, ultimately, the laws of science and nature.

It cannot be emphasized enough that correlations do not prove a cause and effect relationship. Strong correlations can certainly suggest a cause and effect relationship but additional studies specifically designed to prove cause and effect and almost always needed to "seal the deal". Perhaps the most common analytical "trap" or oversight is failure to conduct the definitive experiment when possible to do so. Much of the public as well as many investigators are often too willing to accept a positive correlation as proof of a cause and effect relationship.

Inevitable Correlations:

Unrecognized irrelevant and inevitable correlations, however, represent a significant "trap" that we need to be on the lookout for. These can occur in situations where several different factors or behaviors are correlated with the response of interest and where one of those factors is simply correlated by chance, an irrelevant correlation or where two of those factors or behaviors are unknowingly and necessarily correlated with each other, an inevitable correlation. While most irrelevant correlations are easily spotted, inevitable correlations can be more difficult to discern.

A simple, though unlikely, example of an inevitable correlation might be if an investigator notices a positive correlation between people who have tattoos and an increased incidence of lung cancer. If the correlation is strong enough, say a two fold incidence in lung cancer in people with tattoos, and this is reported in the news, the tattoo business segment of our economy may take a significant hit. However, on further study another correlation is discovered in this population. It's noticed that people with tattoos are twice as likely as the general population to smoke cigarettes. Could it be that the correlation between tattoos and lung cancer is just an inevitable correlation byproduct of the truly meaningful positive correlation between cigarette smoking and lung cancer and the secondary correlation between people who smoke and those who are prone to get tattoos? Hooray! The tattoo artists are back in business. While this inevitable correlation example was obvious, others may be far more subtle and, in fact, may remain hidden for many years giving rise to misleading perceptions of nature.

Related to inevitable correlations are irrelevant chance correlations, i.e., observed correlations that occur strictly by chance but have no particular connection to the matter of interest. For example, it might be observed in a certain study that a high percentage of those who develop diabetes also collect stamps, Since there is no known rationale for suggesting that stamp collecting causes diabetes, this very strong correlation would likely be dismissed as an irrelevant chance correlation. However, it's also possible that other such irrelevant chance correlations may not be quite so obvious and may cloud the issue of focusing on the more relevant correlations.

The key in avoiding the trap of inevitable and irrelevant correlations is, first of all, to be aware of the full range of variables that are also correlated. The investigator should then attempt to justify ruling out all of those likely to be inconsequential in terms of the suspected cause and effect relationship under study. However, in most cases there is more than one correlation that cannot be ruled out without conducting controlled studies.

For example, population studies are frequently used to establish a cause and effect relationship such as the role of dietary fat in incidence of coronary artery disease. One might find that a certain population with a relatively high incidence of coronary artery disease also has a relatively high content of fat in their typical diet. While it's very tempting to claim that such a finding supports the role of dietary fat in the cause of coronary artery disease, the hypothesis really requires further testing where all other variables, such as amount of exercise, smoking, other constituents of the diet, etc. are controlled. Additionally, one should also give equal consideration to certain variables other than dietary fat to see if similar positive correlations can be found in controlled studies. It may well be that one isolated factor such as dietary fat may be insufficient to prove a cause and effect relationship and that the true cause may be a complex of several factors including unknown genetic factors.

For this reason and because controlled experiments are very difficult, expensive and laborious to conduct in humans, such population-based hypotheses seldom achieve the level of scientific fact and we are left with interesting correlations that are suggestive of possible cause and effect relationships but remain unproven. Frequently, different investigators come up with different correlation outcomes and conflicting opinions are expressed. In such cases, further study is needed in order to suggest a rational hypothesis.

Anecdotal Reports:

Anecdotal reports abound in the lay press and there is also a strong tendency to draw sweeping conclusions from seemingly amazing anecdotal reports, sometimes involving very few subjects. This "trap" is likely to snare large segments of the population. Many times such reports are key aspects of advertising campaigns and often center around celebrities. The many testimonials we are constantly exposed to in TV ads often represent modern day versions of the traveling medicine shows and "snake oil salesmen" of years gone by. Pills to give you more pep, more virility or improve strength as well as a plethora of devices and exercise plans to enhance physical appearance and, by inference, increase desirability to the opposite sex are an everyday component of TV viewing. The metallic bracelets, that are claimed to increase balance, strength and overall physical performance, are a good example that we have all seen in a variety of TV ads. The bracelet is often worn by a well-known celebrity, performer or athlete, who attests to the magic that can be had for a modest sum.

With the exception of products that have undergone actual scientific study such as the ethical pharmaceuticals, the ads and infomercials promoting the benefits of these other products are often an obscure mix of fact and fiction. The ads contain enough scientific sounding information to convince prospective customers that there is sound science behind the advertised claims. In fact, many times that perspective is untrue or exaggerated. However, a certain number of vulnerable viewers are "desperate" enough to be misguided into this kind of "scam" purchase, especially if the promoter looks distinguished and is wearing a white coat.

The question is, how can you differentiate between such "scams" and a legitimate product? The ads contain just enough truthful information to escape government false advertising laws and the wording of the exaggerated claims is crafted carefully enough to slip by. By paying close attention to the wording used to describe the claims, most of us should usually be able to tell if the claims are suspicious and likely to be exaggerated. Examples include phrases like "positive results can be expected in up to 80% of those using product X". " SlimAgain contains 30% less fat'. The words "up to" in the first statement and "30% less" in the second statement should represent red flags to the audience. After all, 10% improvement satisfies the "up to 80%" statement, and 30% less than what? The old adage still holds. "If it sounds too good to be true, it probably is not true".

As illustrated above, it's prudent to be especially careful of statements using naked percentages, i.e., percentages that sound impressive but have no real meaning. For example if it's claimed that a certain product has 30% fewer calories, what does this mean? You should always ask 30% fewer than what? Even if it's stated that it has 30% fewer than a previous version of the product or even a competitor's product, you need to ask if that is factual, so what? What is the significance? In another example, a statement may claim that medical data suggests that a certain procedure reduces the incidence of a particular medical problem by a third. However, when you look at the raw numbers you may see that the problem typically only occurs rarely, for example in 3 people per 1000, i.e., 0.3%. The medical procedure is, thus, expected to reduce this incidence to 2 people per 1000 or 0.2%. It's perfectly true that when tested the procedure reduced the incidence of the problem by a third, but ironically that translates to 0.1% of the general population. One needs to ask if that potential benefit is worth the cost of the procedure or, more importantly, the risk of an adverse event caused by the procedure that could potentially occur in a larger subset, for example in 10% of the cases, i.e., 100 per 1000? The lesson here is that when a percentage is touted as an advantage, we usually need to look at the actual numbers behind the percentage.

Chapter 5
Study Design - Prospective Studies

Prospective studies differ from retrospective studies in that study details are specified in advance giving the investigator significantly greater control over the experiment. With this control, the investigator is often able to test the effects of systematically changing one variable while keeping others constant. Thus, any effect elicited during the course of the experiment is quite likely caused by the one variable that is changed.

For example, one could study the effect of relative humidity on evaporation rate from a dish of water in a sealed chamber by holding all variables constant except for relative humidity, which could be varied incrementally, measuring evaporation rate at each relative humidity step. Similarly, the effect of a particular diet on some aspect of health could be studied in laboratory animals with most other relevant variables controlled. These kinds of prospective studies are common in the laboratory setting and used extensively by research scientists in many disciplines.

Prospective studies are used extensively in the pharmaceutical industry in drug discovery programs. Typically new chemical agents are tested for potential therapeutic use in either tightly controlled *in vitro* experiments, i.e. test tube studies, or in live animals, i.e. *in vivo* trials. Since large sums of money in the hundreds of millions of dollars and long periods of time, e.g., 10 years or more may eventually be invested in promising new agents, great care is taken to insure the accuracy of early laboratory investigations.

While such studies are of great commercial and academic interest to the scientific community, they are not generally the type of study that finds its way into the news media or the popular press. Typically, it is the retrospective studies or less well controlled prospective studies that frequently are closer to real life situations and, for that reason, are of greater interest to the general public.

Prospective studies that are less well controlled include human studies, planned in advance, but where all but one variable are not held constant. In many such human studies it is simply impossible to hold all but a single variable constant. For example, one may wish to study the possible benefits of a particular diet on human health. Such a study would likely require a prolonged period where diets are controlled. While subjects may be willing to tolerate strict dietary control for the duration of the study, it is unlikely all other factors that could possibly impact human health, such as exercise, hours of sleep, environment, interpersonal contact, hand washing, etc., etc., could be held constant. Additionally, it is essentially impossible to assure genetic consistency of the large numbers of subjects required in order to obtain statistically significant results.

Furthermore, even if statistically significant results are obtained in certain human trials, they may still be open to question because the impact of uncontrolled variables separately or in combination may be poorly defined. In order to mitigate against the influence of uncontrolled variables the investigator will often attempt to structure the study groups so that they are as similar as possible, i.e. age and sex matched, matched to ethnic background and matched in all other known relevant variables to the extent possible.

The following study is briefly introduced in order to both appreciate the difficulty in conducting a human prospective study and understanding the source of some potential analytic traps and misconceptions. A more complete consideration of this trial is the subject of Chapter 9.

A study reportedly demonstrating the ability of the cholesterol lowering drug pravastatin to reduce the incidence of coronary heart disease in men was reported in the New England Journal of Medicine in 1965. This study, commonly referred to as the "West of Scotland Study", was instrumental in leading to the ongoing widespread use of a whole class of "statin" drugs for prevention of coronary heart disease in millions of people with "elevated" levels of blood cholesterol.

Nortin Hadler, M.D. is representative of a number of those who have questioned the usefulness of pravastatin and other members of the statin class of drugs in preventing coronary heart disease as well as the premise that a "high" level of blood cholesterol is even a major factor in the cause of coronary heart disease (as was mentioned in the Introduction). In commenting on steps taken by the study investigators to assure comparability between placebo and drug-treated groups, Dr. Hadler states in his book "The Last Well Person" (McGill-Queen's University Press, 2009)– "some crucial attributes cannot be measured. Some remain hidden because they are not yet defined (for example, genetic factors that determine collateral vessel growth and the likelihood of healing a myocardial infarct) or because measurement is not feasible. --- The researchers proceed on the assumption that the random assignment of the men will equalize these unmeasured cofounders, or variables. But what if it doesn't."

Further details of the study will be examined in Chapter 9 and will illustrate how difficult it is to reach unanimously accepted conclusions from human prospective trials, even when it is a very large sophisticated study (e.g., >6,500 subjects in the "West of Scotland Study").

Chapter 6

Data Presentation

The manor in which data are presented can have a major impact on how a scientific hypothesis or story is perceived, interpreted and accepted. Data presented in a scientific journal are usually laid out in great detail in a manner that is held to high standards as required by the journal's editorial board. At the other end of the spectrum, data presented in a TV infomercial can be incomplete and manipulated in a way intended to create a contrived impression on the audience. A popular strategy for gaining acceptance of a scientific concept or maybe one that only looks and sounds like a solid scientific concept is to have it presented by our familiar person in a white coat. That simple wardrobe choice immediately endows the wearer with a high level of credibility in the minds of many in the audience. Most likely, the person in the white coat is not a highly trained scientist or physician but rather an actor who has been given the white coat to wear while he or she makes the pitch. Combine this with a clever script that contains a bit of high tech terminology and the audience

59

can hardly wait to phone in their credit card numbers in order to obtain this latest scientific cure that will solve all their health or relationship problems. So, what should one do who hears such a promising miracle from a person in a white coat? The first question to ask is – who is this person? Are they being represented as a physician, dentist or other scientifically trained person? If they are, their title and professional affiliation should be clearly indicated. If not, you can probably assume they are simply an actor in a white coat and judge any opinion they express accordingly.

Experimental results and data presented in the popular press may often lie somewhere in the middle ground. While we may naturally become cautious when we are presented with a miraculous new tonic in a TV ad, more subtle selling of an idea may not arouse the same level of caution when presented to us in the popular press or broadcast news. However, since "science-like" material presented in the popular press has not always undergone critical peer review, we should also use caution here and not immediately accept everything as fact. A major reason for this caution is based on an understanding of human nature. Ideally, an investigator should be totally unbiased in approaching a study. He or she should have no pre-conceived notion of the results he or she hopes to find and no vested interest in the results turning out one way or the other. As you might guess, such investigators are few and far between. Finding anyone totally devoid of bias and free of preconceived ideas regarding any matter may be next to impossible. However, the investigator should at least recognize and be honest about biases that he or she may have. Additionally, reasonable precautions, such as study blinding, should have been taken to counteract these biases and the audience should be so informed.

The form taken in presenting any relevant data should be carefully studied and the question asked as to why the data are being presented in this particular format. We'll now take a look at some of the popular formats for presenting data and explain how these can be constructed to influence conclusions reached.

While it's already been pointed out that caution is warranted when we are not presented with the primary data from a study but rather some derivative of that data such as percentages, we'll now explore how data depicted in graphical form can also be less than totally forthright. The graphical method of presenting data often appeals to the presenter because it can provide an information picture that can easily communicate a point. However, there are a variety of ways to influence conclusions reached by using certain tricks of the trade that are a bit crafty although not entirely unethical.

The first trick originates in the author's selection of what will be represented, the full data set or a selected portion of the data. This choice also may allow the author to custom design the coordinates of his graph to guide the audience's impression of the results in a desired direction. For example, if the author wants to focus on a change in a particular variable with time, it can be magnified graphically by laying out the coordinates to cover only the small ranges of values where the change has taken place. This technique was used very skillfully by some advocates of global warming as seen in the example depicted in Figures 1 and 2 Figure 1 represents an example of a complete data set and Figure 2, a selected portion of data.

Figure 1.

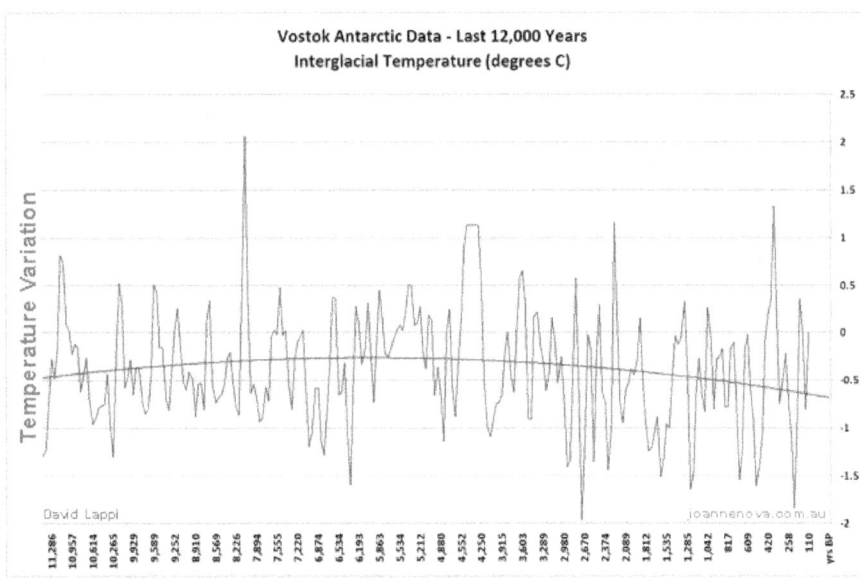

Fig.1 A typical graph of temperature changes in degrees C over the last 12,000 years. This graph depicts the cyclic nature of temperatures that have taken place over a long time span, i.e., the last 12,000 years. Note! Total temperature range is 4 degrees (+2 degrees to -2 degrees). Reprinted from 65 million years of cooling by David Lappi; joannenova.com, 2/18/2010.

Figure 2

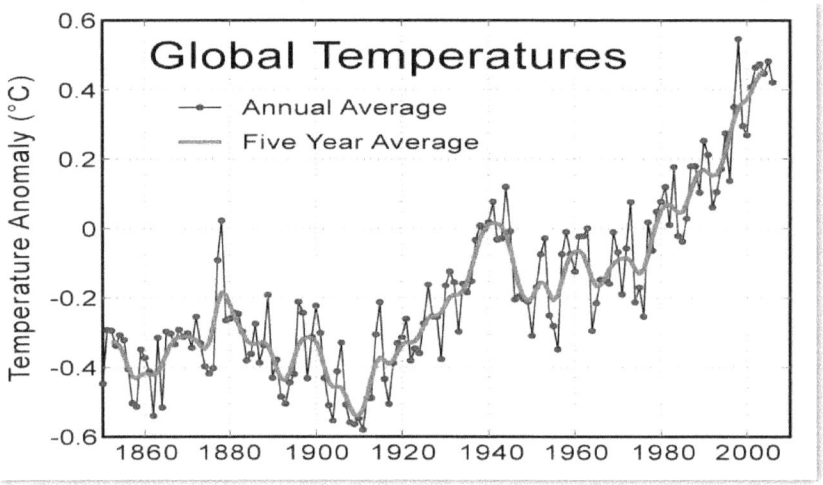

Fig.2 Techniques for creating impression that current upward trend is particularly remarkable. (Note! Temperature range is roughly a quarter of that in above Fig.1 and time span covered in horizontal axis is only slightly more than 1% of that depicted in Fig.1. Notice how Figure 2 is representative of only the last upward spike in Figure 1 and gives no indication that this current upward swing in temperature is characteristic of many former changes as seen in Fig. 1.

Reprinted from figure created by Robert A. Rohde from publicly available data. Global Warming Graphs: L. David Roper, http://arts.bev.net/RoperLDavid/ 24 August 2011.

Even more dubious are graphs depicting an apparent large change in some factor where the author has conveniently failed to fully label the coordinates, sometimes not at all. This may be simply sloppy work but it also might be part of an intentional effort to deceive.

Sometimes histograms or bar graphs are used to represent data. A favorite trick here is to not graph the entire range of variables being presented but only part of the range in order to magnify differences. For example, if a painting's value is increased from $50 to $60 by framing it for sale, that increase would look far more impressive when depicted by a histogram having a scale range from $40 to $60 than by a histogram having a scale range from $0 to $60. In another example, it's also possible for a histogram to give the appearance that something has been reduced dramatically when, in fact, the reduction may be relatively modest, e.g., only from 48% to 37%. One might form a false impression of the significance of the change if a narrow graphical range is represented e.g., from 30% to 50% rather than the full range from 0% to 50%. Another histogram trick is to utilize unequal vertical intervals.

Careful inspection of axes and scale will enable the reader to avoid being fooled by histograms intended to mislead.

Data is also frequently presented in other graphical forms such as pie charts, various pictograms or simply as tabulated data. Aside from the format employed to present data and possible trick used to influence audience impressions, one should also be cognizant of the number of samples or subjects represented, i.e., the n-size of data groups, whether the data were collected in blinded or unblended fashion and whether or not data are being presented as actual measured numbers, percentages or some other derivative of the raw data.

Further examples of the critical importance of data presentation will be illustrated in the chapters to follow. The bottom line is that observed phenomena can be relayed in a variety of ways, often with the objective of creating a pre-determined conclusion in the mind of the recipient. Thus, we need to understand, as well as possible, the objectives of the individual relaying the data to us. Do they have ulterior motives with the intention of "forcing" an erroneous conclusion and how can we best torture the data to get at its true meaning.

Cherry Picking

Cherry Picking is actually the unethical practice of pre-selecting data points to yield a conclusion that the author of the study wishes to disseminate prior to actually conducting the study or presenting the results. For example if the author wanted to publish a study indicating that his newly discovered herbal remedy significantly resulted in reduction of joint pain, he might conduct a trial of perhaps 100 subjects, 40% of whom would report positive results, possibly not significantly different from a placebo effect. He might then present data on 50 subjects, the 40 reporting positive results and 10 having either limited or no positive response. His stated conclusion could be that 80% of the subjects taking the herbal remedy experienced significant reduction in joint pain. The 50 subjects not included in the report were, perhaps, dropped from the analysis because of some claimed disqualification issue, mostly contrived. Perhaps even more common is the practice of simply reporting results on a handful of spectacular cures, many times with raving testimonials. Statements from several subjects claiming to have been miraculously cured of joint pain they had suffered with for years might be all that's needed to generate thousands of dollars in sales. Remember the quote from P.T. Barnum – "there's a sucker born every minute".

Chapter 7
Misleading Headline?

Headlines or lead stories are frequently structured to grab attention and secure the audience. A brief article in "The Telegraph" (19 Nov 2012) is partially reproduced below and presents a case example where a headline was probably used for this purpose but where it may also possibly influence reader opinion, especially for those who only read headlines.

Headline - People who live in tropics more likely to die seven years earlier

People living in the tropics are likely to die more than seven years younger than those in other regions, according to the first findings of a new global research project.

3:02PM GMT 19 Nov 2012

The "State of the Tropics" study, run by 13 institutions across 12 countries, reported that people living in the world's tropical zones in 2010 had an average life expectancy of 64.4 years.

This was 7.7 years less than those living in non-tropical areas, according to the broad-ranging research project, which was initiated by Australia's James Cook University (JCU).

Overall mortality in the region was affected by disease, conflict, poverty and food insecurity, the study said. Investment in social services, such as health and education, as well as access to water, sanitation and medical technology, were also important factors.

According to the report, Central and Southern Africa had the worst adult mortality rates, with 377 in every 1,000 people who live to 15 years old dying before they reach 60.

That compares with an average of 240 in every 1,000 across the tropics and 154 in every 1,000 for the rest of the world.

The study estimates that all continents except Europe and Antarctica are partly in the tropics and 144 nations or territories are either "fully or partly in the tropical region".

The report found that life expectancy in the tropics has increased in the past 60 years, with people living 22.8 years longer than in 1950.

Infant mortality in the tropics also decreased from 161 deaths per 1,000 live births in 1950 to 58 per 1,000 in 2010, though this is still much higher than the 33 per 1,000 rate in the rest of the world.

Source: AFP

Anyone reading only the headline and the bold text might come away with the opinion that it is the tropical climate that is bad for one's health and perhaps they should reconsider a planned vacation to Costa Rica. Reading further one doesn't really get a clear statement of an alternative to climate as the unifying cause of early death. A number of contributing factors such as disease, conflict, poverty and food insecurity were mentioned but no connection was drawn between these factors and the tropical climate.

Could it be that one or more inevitable correlations are lurking here. We can't say because we are given no real data to assess. However, it might be reasonably assumed that poverty rather than climate may be the key unifying cause of early death. Poor people are more prone to contracting disease because of insufficient access to clean living standards. They are more prone to conflict because of their overall frustration with life. They often live on the fringe of malnutrition. Infant mortality is understandably high due to insufficient prenatal knowledge and care of the new born. Thus, it's possible and I would suggest likely that the correlation between tropical climate and early death is an inevitable correlation resulting from the more important correlation of poverty with early death and the fact that there is a disproportionate number of poor people living in the tropics. Why this might be so is a topic for a different series of studies.

The point to be made here is that even though the investigators in this 13 institution study probably fully understand the above point and, more than likely agree with it, the publisher of the "Telegraph" may have judged that the headline pointing out a correlation of early death with a tropical climate would attract more readers than the rather obvious statement that poverty is related to early death. What do you think?

Chapter 8

West of Scotland Pravastatin Study Revisited!

This case demonstrates just how difficult it may be to reach sound conclusions following studies and also illustrates the fact the experts in a given field may often reach diametrically opposite conclusions. Nortin Hadler has reviewed countless medical publications concerning a wide range of clinical trials and has exposed the fact that much of what is considered to be "conventional wisdom" in medicine is not supported by solid data. He discusses the concept of Type II Medical Malpractice and defines it as "doing something to patients very well that was not needed in the first place". This is to be differentiated from Type I Medical Malpractice, which is defined as "medical or surgical performance that is unacceptable. The key point that Hadler makes time and time again in his writings is that much of what is regarded, as the standard of medical practice is not built on a solid foundation of conclusive data. His book "The Last Well Person" (McGill-Queens University Press, 2009) is a compendium that dissects a litany of diagnostic and therapeutic "conventional wisdom" piece by piece.

Hadler provides a table of results from a large clinical trial of the statin drug Pravachol, assessing its effectiveness in heart attack. This table (Table 2.1) after that reported in Hadler's book and excerpted from the original New England Journal of Medicine West of Scotland article (1965) is depicted below. I certainly don't intend to rehash the rather extensive discussion of the results presented at length by Dr. Hadler but I do want to use it to illustrate several points.

Table 2.1

The West of Scotland Pravastatin Study

Outcomes over Over Five Years	Placebo (3,293 men)	Pravastatin (3,302 men)
Non-fatal heart attack	204 (6.5%)	143 (4.6%)
Death by heart attack	52 (1.7)	38 (1.2)
Death by cancer	49 (1.5)	44 (1.3)
Non- cardiovasc. death	62 (1.9)	56 (1.7)
Death by any cause	135 (4.1)	106 (3.2)

Table 2.1 after that presented by Nortin M. Hadler in the book "The Last Well Person" (McGill-Queens University Press 2009).

If we look only at the category of unfortunate subjects who died as a result of a heart attack, we see that there were 52 in the placebo group and 38 in the Pravachol treated group. Those numbers represent 1.7% and 1.2% of the two groups respectively, a difference that was not statistically significant. However, the study investigators found it appropriate to adjust the figures for ten deaths possibly due to heart attack, and the figures changed just enough, 1.9% and 1.3% to become statistically significant and published an article concluding that Pravachol saved lives. Hadler disputes this conclusion, pointing out that the percent of subjects dying in the treated group is only 0.6% lower, a difference that is only barely statistically significant and offers the opinion that this is not clinically meaningful. The implication is that if the study were to be repeated multiple times, the difference, and thus the apparent efficacy of Pravachol, would disappear. So, the question is, who is right?

The larger question might be does the rather small reduction in death by heart attack seen with pravastatin (1.7% in the placebo group down to 1.2% in the treated group, a difference of 0.5% or 0.6% with further massaging of the data) outweigh the cost and risk of adverse events associated with pravastatin usage. In fact, review of the entire statin drug database still leaves unanswered questions yet millions of patients take these drugs on a daily basis. While this may surprise us, it, in fact, is "par for the course" throughout medicine. That may be why the treatment of human disease is referred to as the practice of medicine rather than the science of medicine.

Equally valid cases might be made on both sides of the question. In terms of the grand scheme of biomedical data, the "West of Scotland Study" hardly provides resounding support for having everyone take a daily dose of Pravachol. Long-term treatment with any pharmaceutical agent is bound to lead to adverse side effects in some finite percentage of those treated. On the other hand, you might conclude that 0.6% or 20 of the Pravachol treated group of 3,302 were spared from suffering a fatal heart attack. For those 20 individuals that small statistical result was of great clinical significance. What do you think? If you were a physician, would you prescribe Pravachol for a 48-year-old overweight man with significantly elevated blood cholesterol levels? These are the kinds of questions physicians need to answer on a daily basis, often armed only with the knowledge of unremarkable or inconclusive study results.

Chapter 9
Concluding Remarks

The examples contained in the preceding chapters hopefully provide a bit of insight into the nature of information produced by scientific studies and how the presentation of results can be used or misused to structure the beliefs of the audience. While these examples represent only a very limited introduction to a subject that has been extensively explored elsewhere, they hopefully provide a glimpse of some of the most common traps that can lead to misinformation.

While dissemination of misinformation is occasionally intentional, more often than not, it is unintentional and results from investigator error or failure to be sufficiently critical of his or her own work. Regardless of the author's intent, the recipient of information from all such studies must be ever vigilant, never accepting conclusions on face value and capable of critically evaluating what they are being asked to believe. This is particularly true in cases where the author's conclusions confirm prior opinions or wishful thinking on the part of the audience.

For example, let's say that we really enjoy a glass or two of wine with dinner. We may be inclined to readily accept the results of a study that suggests a certain health benefit of a daily glass or two of wine. However, it is especially in situations like this, where the probability of bias is high and we welcome such validation of our wishes, that we should be most thorough in our efforts to identify potential weakness with the study analysis and conclusions reached.

Naturally, we also need to be particularly cautious in situations where the author's primary goal is to sell us something, either a product in exchange for our money or an idea in exchange for our support. We receive multiple sales pitches for numerous products on a daily basis as we read magazines and newspapers, watch television and work on the internet. Most of these sales pitches are readily dismissed by the more savvy members of the audience but, judging by their prevalence, many must be financially successful, no doubt appealing to a large more receptive population.

Speaking briefly now about politics, politicians have a strong incentive to disseminate questionable information. Their goal is to gain voter support for the next election. As a result, politician's often tell the public what they feel will garner the most votes and the validity of the information or sincerity of the politician's intentions are of secondary importance. Those in the audience who make themselves aware of the major issues of the day and study the relevant facts probably recognize that political rhetoric has a high probability of being untrue or, at best, greatly exaggerated. Unfortunately political literacy, as is the case with scientific literacy, is distressingly low. How many times have we heard "man in the street interviews" where a high percentage of those interviewed were unable to answer even the most basic questions about our government or our economy?

In conclusion, there are two points that are especially relevant and important to keep in mind. The first has to do with the fact that we are all individuals. Most studies provide data on groups. This is especially important when considering health issues. Study results focus on statistical significance of group response, rarely looking at sub-groups and even less likely at individual variations in response. While it is necessary to present group data in order to statistically confirm study findings, as individuals we each need to further ask ourselves the probability that the group finding may or may not hold for us as an individual. A real life example has to do with recommended drug dosing. At a medical conference, it was reported that the recommended dose for a new drug based on clinical trials in 450 patients was 150 mg/day. One of the female physicians in the audience upon hearing the report asked a very relevant question. Since there are potentially significant side effects with this new drug, wouldn't it be more prudent to recommend dosage based on body size since a small women like herself, weighing only 90 pounds, may be only half the size of a large male? A small woman would essentially be getting twice the dose of the large male and, thus, might possibly experience a greater chance of side effects.

In other situations there may be reasons, genetic or issues of susceptibility, known to certain individuals, that would mitigate against using certain classes of drugs. Knowing one's own body and how it might react to various therapeutic measures is critically important in health care matters.

Finally, we'll conclude with the biggy, the "gorilla in the room" so to speak, the nature of reality. A substantial portion of a discussion of reality has been included because of its vital importance to how we perceive and act on new information. A two compartment model of reality has been proposed. Absolute reality is like the pot of gold at the end of the rainbow. Assuming there is such a thing and we are lucky enough to find a bit of it, the increase in our intellectual wealth could be enormous. In fact, even small portions of absolute reality may greatly enrich our lives.

On the other hand, it is our subjective view of reality that we live with from day to day. It is the composite of what we believe. It provides a compass that guides us along life's journey. Regardless of how precious it is to us and regardless of how much we value it, we need to constantly be engaged in refining it to better approach the ideal of absolute reality. There should be no compromise as we explore nature and the material universe. However, with respect to the realm of subjective reality concerning philosophical ideas, including those of politics and religion, we need to realize that each individual has the right to remain free to form his or her own version of subjective reality. It is never appropriate for anyone to attempt to forcibly impose their own subjective reality on another. We can explain our views and attempt to sway others to our particular viewpoint but that should never progress to coercion or force.

Additionally, it's important to emphasize the critical role played by the human mind in structuring the subjective reality concerning our health and well-being. The medical profession has long been aware of a broad spectrum of maladies known as psychosomatic diseases, i.e., disease arising from one's own internal mental anguish rather than from truly physical ailments or external factors such as microorganisms or environmental stresses. Anxiety, depression, impotence, sleep disorders and various gastrointestinal problems are examples of conditions where one's mental state can play a central role in their origin. Consequently, the savvy primary care physician is aware of this and, in fact, realizes that a significant portion of the care he or she finds successful involves treating psychosomatic causes of disease.

On the other side of the coin, the power of a positive mental attitude can be enormous in supplementing conventional medical treatment and curing disease. The placebo effect that complicates interpretation of most trials of pharmaceutical agents is likely also a result of positive expectations. Faith healers, witch doctors and shamans have long had surprising success. in tapping this potential. Because of the intense mental power brought on by faith, much of this is closely associated with religion. Many of us have experienced a rapid recovery from sickness brought on by a sudden boost in positive mental attitude. Perhaps this has resulted from reassuring words of a physician or by the surprise visit by an old friend. The 1952 book "The Power of Positive Thinking" by Dr.

Norman Vincent Peel was one of the early attempts to successfully convey this message to the general public. Whatever complex mechanism is involved, we can be sure that the human brain plays a central role and, thus,, has tremendous power in structuring our subjective realities. This example alone should leave little doubt as to the importance of our subjective realities as we journey along the pathway of life.

About The Author:

J. Stuart Fleming, Ph.D. completed a 35 year career with a major pharmaceutical company where he served in a variety of capacities from leading both preclinical and clinical research to managing project life cycles from discovery through post-marketing. He earned a Ph.D. in Physiology from Ohio State University after obtaining an M.S. degree from the University of Buffalo and a B.A. from Northwestern University. Most recently he completed an M.B.A. at Syracuse University with a major interest in Innovation Management. He has authored numerous scientific publications as well as the recent self-help book, **A Path To The Gold** and a collection of essays featuring a discussion of Reality in the book, **Views From The Rabbit Hole.**